Military Animals

ANIMALS IN MILITARY ACTION

by Debbie Vilardi

abdobooks.com

Published by Pop!, a division of ABDO, PO Box 398166, Minneapolis, Minnesota 55439. Copyright ©2022 by Abdo Consulting Group, Inc. International copyrights reserved in all countries. No part of this book may be reproduced in any form without written permission from the publisher. DiscoverRoo™ is a trademark and logo of Pop!.

Printed in the United States of America, North Mankato, Minnesota.

102021
012022

THIS BOOK CONTAINS RECYCLED MATERIALS

Cover Photo: US Department of Defense
Interior Photos: US Department of Defense, 1, 20, 23; Defense Visual Information Distribution Service, 5, 6, 7, 14, 16 (right), 17, 19, 25, 26, 29; Shutterstock Images, 8–9; North Wind Picture Archives, 11, 16 (left); AP Images, 13; Sam McNeil/AP Images, 15; iStockphoto, 27

Editor: Charly Haley
Series Designer: Laura Graphenteen

Library of Congress Control Number: 2020948914
Publisher's Cataloging-in-Publication Data
Names: Vilardi, Debbie, author.
Title: Animals in military action / by Debbie Vilardi
Description: Minneapolis, Minnesota : Pop!, 2022 | Series: Military animals | Includes online resources and index.
Identifiers: ISBN 9781532169939 (lib. bdg.) | ISBN 9781644945889 (pbk.) | ISBN 9781098240868 (ebook)
Subjects: LCSH: Animals--Juvenile literature. | Working animals--Juvenile literature. | Military readiness--Juvenile literature. | Armed Forces--Juvenile literature.
Classification: DDC 355.424--dc23

WELCOME TO DiscoverRoo!

Pop open this book and you'll find QR codes loaded with information, so you can learn even more!

Scan this code* and others like it while you read, or visit the website below to make this book pop!

popbooksonline.com/military-action

*Scanning QR codes requires a web-enabled smart device with a QR code reader app and a camera.

TABLE OF CONTENTS

CHAPTER 1
Horses in Combat 4

CHAPTER 2
A History of Service 10

CHAPTER 3
War Dogs 18

CHAPTER 4
Other Animals in Action. 24

Making Connections. 30
Glossary 31
Index. 32
Online Resources 32

CHAPTER 1
HORSES IN COMBAT

Three horses walked on rough ground.

They carried American soldiers in

Afghanistan in 2001. It was the first

US military attack using horses since 1942.

Some of the soldiers had just learned

to ride horses.

WATCH A VIDEO HERE!

A US Army soldier rides a horse while training to work in Afghanistan.

American soldiers bought horses from farmers in Afghanistan.

The horses could go where vehicles could not, and few vehicles were available. The American soldiers' goal was to force dangerous enemies out of the city of Mazar-e-Sharif.

A US Army soldier feeds a horse.

1

Horses have been used in **combat** for 3,000 years. Other animals work with militaries too. Animals sniff out **mines** or enemy soldiers. They find people who are wounded and help them. They warn soldiers about attacks. Some animals patrol borders and camps.

A US military dog searches for hidden weapons in Iraq.

DID YOU KNOW? World War I (1914–1918) soldiers read maps and letters by the light of glowworms.

9

CHAPTER 2
A HISTORY OF SERVICE

Different animals have been used in wars. Elephant **cavalries** date back to between 400 and 301 BCE. A military general named Hannibal used elephants against the **ancient** Romans.

LEARN MORE HERE!

Hannibal, his soldiers, and the elephants made a dangerous journey through the European Alps. This path was sneaky and surprised the Romans.

Hannibal used snakes in battle too. His navy **catapulted** jars of snakes onto enemy boats.

Soldiers ride an armored elephant in ancient Rome.

War changed in the 1900s.

During World War I, soldiers hid in **trenches**. Poison gas was often used in battle. It hurt people and animals.

To protect themselves, American soldiers brought slugs into trenches. Slugs stop breathing when gas is in the air. When they're not breathing, slugs

> **DID YOU KNOW?**
> In ancient Rome, some armies used pigs to fight against war elephants. The elephants were scared of the pigs and ran away.

During World War I, American soldiers fought in trenches in France.

look smaller. Soldiers watched the slugs to know when it was time to put on gas masks.

Many other animals still help militaries today. Jordan's military uses camels to catch enemies. In the 1960s, the US military began training dolphins.

A US Navy dolphin listens to instructions.

Members of the Jordan military ride camels during a parade.

Dolphins are still used to find enemy swimmers today. Many militaries also use dogs. Dogs can sniff out enemy soldiers or weapons. They have been used since 600 BCE!

TIMELINE

1942
The US military starts the War Dog Program to train dog soldiers. The military had used dogs in battle before.

1500 BCE
Soldiers begin using horses in battle.

1914–1918
Many animals served during World War I. Cats hunted mice on ships and in trenches. Slugs warned soldiers of poison gas.

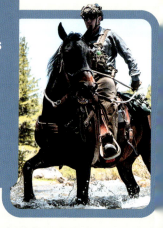

1965
During the Vietnam War (1954–1975), Vietnamese soldiers throw beehives at American soldiers and place other hives along marching trails.

1960
The US Navy begins the Marine Mammal Program. It trains dolphins, whales, and sea lions.

2000s
Scientists study different ways that militaries could someday use insects.

WAR DOGS

Military dogs are very useful. They are brave, loyal, and smart. They have strong senses of smell. They hear things people can't. Most militaries use dogs today.

COMPLETE AN ACTIVITY HERE!

Military dogs are trained to attack on command.

Handlers spend time caring for their dogs.

Handlers train the dogs. They give commands such as "sit" or "stay." But handlers and dogs may need to stay quiet in military action. So handlers also use hand signals to talk to dogs. Handlers reward the dogs with petting or play.

HERO DOG

Stubby was a dog that fought for the United States during World War I. Stubby warned soldiers when he smelled enemy gas attacks. He also helped wounded soldiers. Once, Stubby captured an enemy soldier. This earned him the rank of sergeant. Sergeant Stubby was wounded in battle, but he had surgery and healed. He came home a hero.

Military dogs do many jobs. They sniff for bombs. They listen for enemy soldiers. They help protect military camps. Dogs also find soldiers or other people who are in trouble. They may even jump from planes to get to where they are needed!

A dog's work may be dangerous. Goggles protect its eyes from sand. Some dogs wear body armor. They may also wear gas masks.

DID YOU KNOW? **The US Postal Service issued stamps to honor military working dogs. The four stamps came out in 2019.**

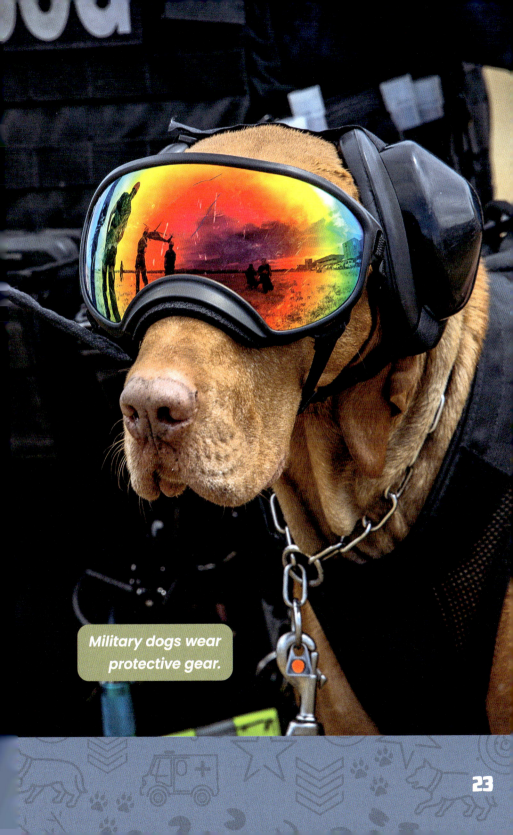

Military dogs wear protective gear.

CHAPTER 4

OTHER ANIMALS IN ACTION

Other animals serve in the military too. The US Navy has trained dolphins and sea lions. These animals are smart. They can be easy to train. They also have skills that humans do not.

LEARN MORE HERE!

A US Navy sea lion and its handler salute each other.

Sea lions see well underwater.

They find lost items. They may attach lines to help sailors pull the items up.

They can also find enemy swimmers.

Dolphins have **sonar**. They seek out **mines** and enemy swimmers. They mark the spots with tags. The animals return to their ships. Then their handlers attack the enemies or remove the mines.

Military members watch as a dolphin heads into the water to find objects during training.

DOLPHIN SONAR

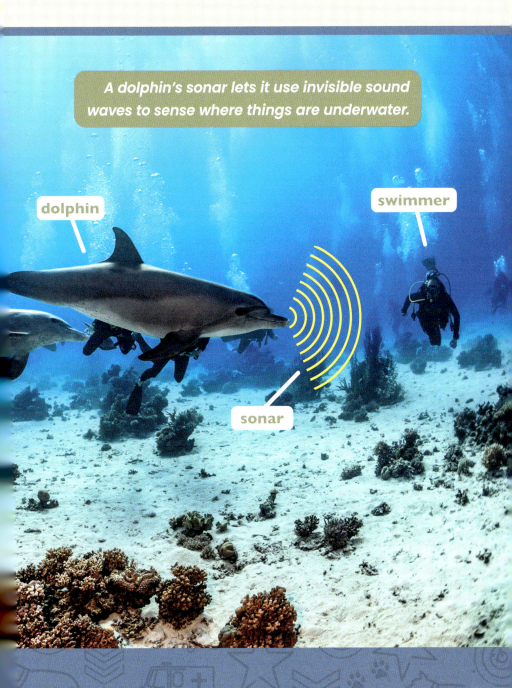

Insects are animals that could help militaries in the future. Scientists are studying insects such as moths and cockroaches. They want the insects to fly or crawl where they are told to go. They may carry cameras. They may also spy or look for hurt people who need help.

DID YOU KNOW? Scientists are making robots that are similar to animals. These include robot lobsters. They could be used to find mines in the sea.

Members of the military study insects that carry diseases.

MAKING CONNECTIONS

TEXT-TO-SELF

Would you allow your pet or favorite animal to help soldiers in a war? Why or why not?

TEXT-TO-TEXT

Have you read other books about animals in the military? What were their jobs? How did they help people?

TEXT-TO-WORLD

The animals in this book help people in militaries. How else do animals help people? How can you help animals?

GLOSSARY

ancient — from a long time ago.

catapult — to use a weapon that throws large stones or other objects through the air.

cavalry — a part of an army mounted on animals.

combat — fighting between enemies in war.

mine — an explosive weapon often put underground or underwater.

sonar — in biology, the method of echolocation used in air or water by animals such as bats or dolphins. Animals do this to learn about where they are and what is around them by making sounds and listening to how they bounce back.

trenches — deep, narrow pits that many soldiers fought from and hid in during World War I.

INDEX

camels, 14
cavalries, 10

dogs, 15, 16, 18–22
dolphins, 14–15, 17, 24, 26, 27

elephants, 10–12

Hannibal, 10–11
horses, 4, 6, 9, 16

mines, 9, 26, 28

sea lions, 17, 24–25
snakes, 11

trenches, 12, 16

ONLINE RESOURCES
popbooksonline.com

Scan this code* and others like it while you read, or visit the website below to make this book pop!

popbooksonline.com/military-action

*Scanning QR codes requires a web-enabled smart device with a QR code reader app and a camera.